SHADOW PEOPLE —THE THREAT OF DARKNESS

By

Rob Shelsky

~DEDICATION~

For:

Diane Powell and George Kempland

Friends of the Best Sort

Always and Forever!

Contents

Introduction

A grown man, such as myself, seriously mystified and frightened by a mere shadow? In my case, this is exactly what happened. My living room and kitchen in my San Diego home were united in an open floor plan. From the kitchen counter I could look directly out at my front yard or rather, onto my front deck. This was entirely enclosed to a height of eight feet by a solid fence. On this fence, in a number of horizontal rows crossing the entire structure, I had stag horn ferns mounted, dozens of them.

One afternoon, I was cutting up vegetables for dinner on the kitchen counter's cutting board when a movement caught my attention out of the corner of my eye. I glanced up. A shadowy figure, which seemed male in general aspects, moved swiftly across my front deck. It was amazingly swift I must add, faster than a human could possibly move. The window, which was ceiling-to-floor, wall-to-wall, plate glass gave me an unobstructed view of the entire front deck. Again, the thing I saw was that of a shadowy-like figure, perhaps male in general nature, but lacking in all details. It was there, and then gone in an instant!

Surprised and alarmed, I thought someone must be out the front of the house seeking a way in, perhaps a burglar or some other type of intruder. This, despite the fact that I was sure I had locked the eight-foot gate (this was to keep potential salesmen or other types of solicitors from bothering me at the front door—a common nuisance in San Diego at the time). Had someone somehow opened the gate? Had I been wrong about having locked it?

Confused, I set down the vegetable knife on the counter (in retrospect, I should probably have taken it with me for protection, but I didn't) and went to check out the situation. I opened the door and stared out at the small deck area enclosed by the fence.

Nothing.

No one was there. I walked out and checked the gate to see if it was locked.

It was.

No one could have entered. The gate was most definitely secured shut.

Even so, the tiered rows of staghorn ferns along the interior fence wall all wafted back and forth. It was as if someone had just brushed closely by them, their fronds waving about, although the air was completely still. There was no sign of any breeze.

Yet, they were moving just as if someone's passing had set them in motion. I searched around the outside of the house, front and back, still thinking somehow an intruder had been there, had somehow gained entry, although to scale the sheer fencing, again, some eight feet in height, would have been difficult, to say the very least. Despite my extended search, again, there was nothing unusual to be seen. There was no one.

Needless to say, this not only bewildered and confused me, but it shook me up, as well. I wasn't at all sure of just what I had witnessed. But I was sure of one thing; I had seen *something!* So convinced was I of having seen that "something," that I did some research. I even went so far as to check with the prior owners, the ones I had directly purchased the house from, to see if they had seen any signs of the home ever having been haunted when they had owned it. Yes, I was that unnerved by the experience!

To go to such lengths sounds extreme, I know, but again, by that point I was grasping at straws to find some sort of answer. The prior owners were adamant that the place had not exhibited any such thing as apparitions of any sort, let alone ghosts. Yet, despite all this, I was certain I had seen something. I was utterly convinced of this.

This was my first of about three major personal experiences with what I considered bona fide shadow-person sightings. There were other less extraordinary ones, but this had been the major sighting for me. Three points to remember with this story:

1. I definitely saw a dark shadowy figure flit across my front porch. This was not my imagination, not a trick of the lighting, and not an illusion. I would swear to this under oath in a court of law.

2. The figure caused the many staghorn ferns I had on the fence/wall there to sway markedly. So there had to have been some sort of presence, something or someone with some sort of physicality, enough to move air aside as they passed.

3. "Something" had to have caused this. That cause wasn't the wind, and it wasn't me. I don't know how it could have been anyone else. So, what was it? What was left that might have accounted for this movement of all my plants? The answer is, of course, there was no other "logical cause" that I could think of. So, what was left, on the face of it, seemed illogical to me, but still true, nevertheless.

4. Furthermore, there had been an unaccountable and deep; I would have to say, almost primeval fear, which filled me at the time of the sighting. I'm not normally or easily frightened. I usually seek alternative explanations for bizarre things happening, more reasonable, everyday ones. I do consider myself a rational human being. However, this time I could not find any rational alternative explanation. Moreover, my fear seemed unnatural for the scope of the event, distinctly (even to me) out of proportion to what had occurred. It was a visceral sort of fear, the kind one might feel when entering a dark cellar and hearing something unnatural moving about below, in the dark depths of the stairway. Yeah, it was that strong.

When I have dared to relate the experience on various occasions to different people (something, I quickly learned to stop doing), they have always tried to explain it away to me as being just a case of me mistaking something "normal" for something worse, perhaps some sort of mental aberration, but again, *I know what I saw*. And truly, I had felt that instant and powerful pervading sense of fear at the time. That same fear is what directly motivated me to go and check everything out so thoroughly after the event, to make sure there was no burglar or intruder of any sort lurking on or about the premises. Once more, I had even checked to see if the place might have a history of being haunted! I had desperately wanted some sort of answer as to what had caused the event, and exactly what the nature of it had been. I didn't know what to think.

I was at a loss. My thoughts went round and round. With a high and solid fence and a locked gate, I don't know how anyone could have gotten in, but whatever this had been, the fence had seemed like a vapor to it, for it had to have passed right through it to get out...apparently. It had simply flitted across the front decking from left to right and then had utterly vanished.

Nor am I alone in such sightings. There are literally tens and even hundreds of thousands of such things reported and constantly. Instances that have been caught on camera are all over the Internet, ones verified to have been legitimate, unretouched videos. Moreover, these sightings are not restricted to any one region. They occur worldwide.

Shadow People, as they are called, are seen in rural houses, city apartments, alleys, streets, woods, parks, babies' bedrooms, hotels, onboard ships– you name it – they've been seen practically everywhere. Responses to such visitations have often included surprise, shock, and yes, on many occasions even the same visceral sort of fear that I had felt.

Sometimes, children are seen playing with or talking to shadowy figures lurking in the corners of their room. Some children don't seem afraid of them, but others will scream and cry in panic when confronted by such things. At the end of this book, I will leave a series of links to various videos of various sightings on YouTube. Do check them out. Some are VERY compelling examples of just what I'm talking about here. Moreover, they will help allow you to decide for yourself if Shadow People are real or not.

Conclusion: The Purpose and Scope of this Book.

So, what are Shadow People? Do they even exist? If so, who or what are they? Why are they here? What do they want? And if they do exist, are they a danger, a real threat to us? In the next chapters, we will try to find answers to these questions. Because, whatever Shadow People are, if they exist, then what they may portend for humans and the future of humanity might just be the biggest mystery we need to solve if we are to survive.

Chapter 1—What are Shadow People?

For those who may be wondering what a shadow person is, in the modern sense of the term, and would like a definition, here is how Wikipedia puts it:

"The first time the topic of Shadow People was discussed at length...and first came to prominence was on Art Bell's radio show, Midnight in the Desert. On the show **[dated]** *April 12, 2001, when host Art Bell interviewed a man purporting to be a Native American elder, Thunder Strikes, who is also known as Harley "SwiftDeer" Reagan. During the show, listeners were encouraged to submit drawings of Shadow People that they had seen and a large number of these drawings were immediately shared publicly on the website."* **[Emphasis added.]**

But that was just the beginning. Interest in these seemingly nonphysical creatures, which often appear to have vaguely human shapes, but not always, continued to grow, and as Wikipedia also says:

"In October that year, Heidi Hollis published her first book on the topic of Shadow People, and later became a regular guest on Coast-to-Coast Radio with host George Noory. Hollis described Shadow People as **[being]** *dark silhouettes with human shapes and profiles that flicker in and out of peripheral vision, and claims that people have reported the figures attempting to' jump on their chest and choke them.' She believes the figures to be negative aliens that can be repelled by various means, including invoking "the Name of Jesus".* " **[Emphasis added.]**

Again, this is according to Wikipedia. And, as usual with such types of topics, especially on such controversial radio shows, the consensus of what Shadow People actually are soon began to break down as different people with differing beliefs became involved. But Ms. Hollis' definition, for the most part, seems to be a good working one for our purposes here in this book…to a point. I'm not at all sure that religious implications need to enter into such explanations and descriptions, but neither do I preclude them entirely, either. One must keep an open mind when it comes to considering such seemingly paranormal matters. But as always, I think we should first try to find answers in our physical world before we resort to the metaphysical.

After all, who hasn't seen something fleetingly out of the corners of their eyes, seen a brief dark shadow that shouldn't be there, that was there for an instant and then gone? Again, one brief second, it was there and the next moment it was gone. This usually occurs at odd moments and often when we least expect it. Whether this happens while we're sitting and watching TV, reading a book, or simply standing or moving about the house or outside, we might catch a momentary glimpse of "something" at the very edge of our vision.

Again, it was something that seemed to be there and then just as swiftly was gone. Many times, that "something" could disturb us greatly. Quite often, there is sense of consternation, curiosity, innocent bewilderment, and many times as mentioned above, even some sort of dread or fear ensuing as a result of such sightings. This often results when the observer is certain that there had been something there and that it wasn't just "a trick of the lighting," an "illusion," or a simple mistake on the part of the viewer.

However, and this is notable: there are exceptions to how very brief these sightings can be. Occasionally, whatever it is, does remain long enough for people to even get a picture or short video of the event. Moreover, again, sometimes, that "something" can be very frightening, even terrifying in appearance and its implications for us as humans.

However, we will get more into that aspect of things a little later in this book. Suffice it to say, as a result of being captured on video and in photographs, one can easily find such on the Internet. Could some of these photos and even videos be hoaxes? Yes, some most certainly are. I have no doubt of this. But only some, not all, are fakes. Some are posted by reliable, down-to-earth and reputable witnesses of such events. Furthermore, they have been subjected to analysis by experts in many cases and have been found to be the actual thing, not something that has been photoshopped or had C.G.I. added.

In any case, virtually everyone has at one time or another seen such things, at least in quick glimpses. Often, this has happened, if not frequently to an individual, at least it has probably occurred more often than once. For some, it is a truly oft-repeated event, perhaps even worryingly so, for the fact of the numerous times it has occurred. For others, such events may be rare and perhaps at times, even nonexistent. Again, I stress that nearly everyone has had some such experience during the course of their lives, and often more than once, but there are a very few they may not have.

In the daytime, we construe such momentary glimpses as just an aberration of our sight or a trick of our minds or the lighting, just a simple case of our imagining something was there when it was not. Almost always and invariably, we dismiss such daytime, momentary and transitory events as nothing more than a case of misinterpretation, mistaken identity of something else, or again, so just our imaginations making something out of nothing.

We tell ourselves we are seeing things that simply aren't there. Often, that is probably the actual case. But not always, not in every instance! And more, when we are certain we saw something but then later mentally dismiss it as non-important or just a "mistake" are we intentionally lying to ourselves? And if so, why? Well, one reason might just be...fear! More on that later.

More often at night than in the daytime, we see what some people refer to as an actual shadow person. Shadow persons, also often referred to as shadow figures, seem to appear much more often at night, but again, not always. They can appear in sudden and frightening moments and totally in unexpected situations. Sometimes they are not even a figure, so much as just a sort of black amorphous mass or blob-like entity that moves, often seemingly of its own volition. They can even be just a regular patch of shadow that suddenly takes a specific form (or not) and darts around, although there is no one present to be casting such a moving shadow. In any case, the shadow often seems to be an entity of some sort in its own right. One in a while, Shadow People can look transparent, or mostly so. They flit by as a blurring of what one sees beyond them, as if they are almost, but not quite, invisible.

People who witness such sightings are often convinced they are seeing some type of a creature of some sort, whether of supernatural origin or otherwise, (although most see them as supernatural in nature). They often can see something that resembles a figure, again, one frequently that is vaguely humanoid in appearance. How this image is interpreted by the viewer is, of course, often subjective, yet the basic principles remain. The figure frequently appears humanoid-like, but shadowy, and moves astonishingly fast or simply seems to suddenly disappear. On occasion, they appear to be able to move through solid objects, like walls or furniture. The shadow figures also seem insubstantial and even transparent in many cases.

One other thing should be mentioned; invariably, and without exception, these Shadow People are silent. They make no noise as they move and they utter no sounds. The are silent creatures of darkness, it would seem.

As for explanations? Well, some might see such as a thing as being the manifestation of a ghost or spirit, while others might see it as some sort of being from "somewhere else." The deeply religious might interpret them as being demons or other things of an evil nature. This is especially true of those who deeply believe in the afterlife, the paranormal, the occult or supernatural in general. For them, such entities or eerie creatures are from some other realm. They may not be entirely wrong in this idea because, perhaps, these sightings are of things that if they are not from a different realm, then maybe they are part of a greater realm than we are aware of. Again, I'll explain more on this idea later on, when we discuss this subject in more detail.

Although many commentators on the Internet, ones at sites covering such paranormal topics, think that Shadow People are dangerous, others do not agree with this idea. They strongly debate whether Shadow People are intrinsically evil, or whether they may instead, just be helpful, or even nothing of either sort, but just something very different. These last, are people who have a more neutral approach, with some believing Shadow People might just be extra-dimensional inhabitants that have somehow stumbled into our world. Such people believe Shadow People are just beings from parallel universes, dimensions, or other "realities," but not from any sort of afterlife.

Chapter 2—Are Shadow People Real?

Now that we have an idea of how Shadow People can appear to us, the next question is: are Shadow People even real? Do they actually exist? Alternatively, are they just the stuff of urban legends, nightmares, and myths? If they are real, how long have they been here? Another question is: why are they here?

Are Shadow People real?

Religions seem to believe in Shadow People as being real: A great number of religions, legends, and various related belief systems describe supernatural entities such as shades ("Shadow People?"), perhaps, as having come from the afterlife or are creatures of some version of an underworld. Either way, various types of shadowy creatures have long been a staple of folklore and ghost stories, myths, and legends, and certainly not least of all, religions. These are worldwide in their scope and variety and are everywhere. Such tales date back thousands of years and include things, such as the stories of the Islamic Jinn of the Middle East and the Choctaw, Nalusa Chito, of the Americas. There are many, many more.

I mentioned earlier that the *Coast-to-Coast AM* late-night, radio talk show helped to propagate information about Shadow People. Specifically, the first time the subject of Shadow People was broached in any real depth seems to have been on the April 12, 2001, show. Host Art Bell interviewed a man claiming to be a Native American elder. Whether he really was or not, as of writing this book, is still undetermined by me, but he went by his indigenous name of Thunder Strikes. He also has been called Harley "SwiftDeer" Reagan.

The show then encouraged people to send in illustrations of Shadow People as seen by them. A slew of such hand drawn illustrations was sent to the show. These drawings helped to clarify just how Shadow People looked to those who witnessed them. There were highly consistent themes to these illustrations in that the pictures all depicted the same basic versions of Shadow People as described in the prior chapter. One person who described such entities did so in revealing detail.

In late 2001, Heidi Hollis fortuitously came out with her first book on the idea of Shadow People. Her timing was superb in this regard, as this enabled her to become a frequent guest on *Coast-to-Coast Radio*. In her book, Hollis claims Shadow People often appear as just blank silhouettes, dark shadowy figures with human or vaguely human shapes. Sometimes, they seem to be just shapeless blobs.

These entities often are seen as just a "flash of something dark," a mere flicker that comes into range of our peripheral vision and then, just as abruptly, disappears. Some people have even claimed that in the dark of night, while in bed, such shadows have tried to "jump on their chests and choke them."

Heidi Hollis was convinced (according to her book, at least,) that such Shadow People are "negative aliens "and that they "can be repelled by various means," including invoking "the Name of Jesus." Whether this last is accurate or not is an open question, since it is Ms. Hollis' own conclusion and there seems little to no factual evidence either to support or oppose such claims. Then, gathering information about the subject of Shadow People is no easy task, as I've found out.

In many cases of Shadow People sightings, a person can detect the outline, the shadowy shape of some creature, but there hardly ever seems to be any detail as to regard to specific features of Shadow People, including a lack of facial features, except perhaps (in some few cases) "glowing or red eyes."

However, as harmless as such sightings appear to be in many cases, the observers of such phenomena do more often than not do describe a sense or feeling that these beings, entities, or whatever, are malevolent in nature. Various observers, again, have managed to take photographic evidence of these creatures. For instance, Chad Stambaugh, a researcher, states that he has multiple videos of Shadow People.

Chapter Conclusion: So, do we have definitive proof that Shadow People actually exist? Is there a "smoking gun" to prove their existence. Well, no, there is not, at least, not for some skeptics. However, it seems there are numerous photos, videos, and a mountain of evidence in anecdotal form with regard to such things being real. There is a long and worldwide history of such sightings, as well. One which seems to date back thousands of years.

Moreover, there is a strange allure, a fascination with the idea of Shadow People in pop culture. So not only has the concept of Shadow People, in one form or another, existed for millennia and again, is worldwide in scope, but even today, right now, it grips the imagination and fears of many people. We delve more into this subject a little later on in this book.

Why is there this fascination with Shadow People? What is the allure? We will discuss this in more detail in the following chapter, as well. But based on witness testimony by tens of thousands, and hundreds, if not thousands of photos and videos, many of which have been analyzed by experts, it would seem Shadow People very well may exist!

And to answer your unspoken question, one which I'm sure you have, at least, have mentally asked:

Do I, as the author of this book, believe Shadow People are real?

The answer is yes.

I think they really exist and I find that existence not only strange to my way of thinking as to what is reality, but perhaps the idea of such creatures being real as a possible very dangerous things for us humans and our future, as well.

Chapter 3—The Odd Allure of Shadow People

So just what is the attraction to there being Shadow People, of even investing any energy into the belief in their existence? Well, there are a number of reasons, one of which is based on people's past personal experiences with such things.

1. People having had some sort of personal encounter with them or knowing someone whom they trust that has had some sort of experience with Shadow People often makes those involved want to know more, if only out of curiosity. I originally fell into that category. Having been startled, shaken, and yes, even made fearful about my own sighting of a Shadow Person, I became gradually more and more interested in the subject over time. This then, evolved into researching the subject, including what was currently known, as well as historical evidence and folklore regarding the topic of such entities. Many people start their interest in Shadow People in this way.

Such interest can either grow or wither over time, depending on the severity or impact of our personal experience with such creatures might have been. Some people actually become totally obsessed with the idea of Shadow People and so become infatuated with the idea of proving the existence of such beings. This, for me, is a mistake.

To become so fixated, so adamant about trying to prove something, rather than just trying to solve something makes for clouded judgment in my opinion. Such people would be better off to try to prove or alternatively, "disprove" their existence in an evenhanded way, rather than becoming fixated on just trying to "prove" such things exist (whether they really do, or not). One has to keep an open mind in both directions, attempt to follow the scientific method in such cases or risk having unreliable results, and so perhaps jumping to erroneous conclusions based on such bad results.

2. There is also another attraction, an odd one on the face of it, and that is the "sense of fear." Fear, a sense of danger, or malevolence, which many people experience when viewing such an entity, even though in most cases, the shadows don't do anything dangerous or overtly threatening in their behavior (most of the times) is another factor which seems to cause an allure with regard to the subject of Shadow

People. This is the same sort of allure that exploring a haunted house has for many, or riding some thrill ride at a carnival. We get a vicarious thrill delving into knowing more about Shadow People, as well, as trying to assess if such entities are really dangerous, or a threat to us or our loved ones. And as often happens, such an interest can grow over time, as it did in my case and again, so too, with many others of like minds.

3. Furthermore, the whole idea of shadows and Shadow People has always been fascinating. As one website put it:

"The allure of shadows lies in their ability to conceal, reveal, and transform, making them a fascinating subject of exploration." **[https://fastercapital.com/content/Mysterious—The-Mysterious-Allure-of-Shadows—Unlocking-Secrets.html#The-Enigmatic-World-of-Shadows] [Emphasis added.]**

As that same site also states:

"Shadows also hold a psychological significance, evoking a range of emotions and associations within us. They can elicit feelings of mystery, fear, or even comfort, depending on the context in which they appear. In literature and film, shadows are often employed to create an

atmosphere of suspense and intrigue. Think of classic film noir, where the protagonist lurks in the shadows, adding an air of danger and ambiguity to the narrative."

So, (1) prior experiences with Shadow People, the allure of danger and fear that such sightings can often evoke in the observer, and (2) just a general urge to want to know more, curiosity if you will, with regard to such a phenomenon seem to be the primary causes of people's interest in this subject. Also (3) The air of mysteriousness around it all also acts as a seeming psychological inducement for the rising interest in Shadow People. It seems people just love a good mystery! Most of us do, at least to some degree, but I wonder if the same would hold true if the answers to the mystery were not good, were not ones that people really wanted to know. In such a case, would you want to know? If the answer to this question is no, you would rather not know, then perhaps you should not continue reading this book, for some of the conclusions here are highly disturbing if true.

What do we know about Shadow People, and what they might be doing here? Well, first we have to consider the origin of the idea of Shadow People. From where does this belief in such a thing originally stem? Is there a common source for the wide variety of beliefs regarding such entities that exist today? Why has it persisted for so long? To understand this, we have to backtrack just a bit here to figure that part out. The best way to do that is to gather any information we can about the history and folklore of Shadow People. As the old saying goes, "to know where we are going, it helps to know where we are coming from."

Chapter 4—History and Folklore

The idea of Shadow People, in some version or other, having existed for ages in our cultures is not a new one. History abounds with people claiming to have had personal experiences encountering such entities. Regardless of the culture there seems to be a long, often anecdotal or verbal history of Shadow People. And this dates back at least for several millennia, and probably much longer.

Origins of Shadow People in various cultures. Relying on strictly anecdotal stories, call them legends, myths, or what have you, can often be misleading and make people come to sometimes erroneous conclusions. This is because, over time, verbal legends/myths/stories alter and change with retelling. At least, that true in many cases. Moreover, such legends and myths can be regional and **sometimes** even just local to one small area, so they cannot be used as reliable data, unless the story repeats elsewhere **in the world** and more than just once in a rare while.

Therefore, when it comes to the handing down of stories as history of real events, one often has a hard time separating fact from fiction. However, despite all the telling and retelling there are certain stories that defy being just local or highly regional, and instead are geographically widespread. They may differ in some **minor** details from region to region, but overall, they remain remarkably similar. **The then, are more reliable than just a one-off and seldom repeated local legend.**

Here are a few examples from different cultural traditions in this regard:

1. Native American Folklore:

In some Native American tribes, there are legends of shadow beings or "The Shadow People." These entities are often associated with the spirit world and often they are believed to be messengers or guardians. They might appear as shadowy and dark figures, and sometimes they take on the shape of animals or humans, if not all the detailed aspects of them. The Choctaw Tribe is just one example of a culture that incorporates such beliefs.

The Choctaw, are an ancient indigenous people that have for 4,000 years thrived in North America. During that time, they became quite advanced in some ways, to the point where western Europeans classified them as being one of the "five civilized tribes" because of this. Despite this advancement with regard to their style of government, use of tools, etc., the Choctaw had their Shadow People problems, as well, it seems. One of their stranger beliefs if that of the Nulusa Chito, a shadowy figure that can be manifested by negative thoughts. And like so many other legends of different cultures, these creatures attack people at night, while sleeping, and only they do they absorb the victim's soul. Sound familiar? It should. This could even pass for a description of vampires. The Choctaws other monster was the so-called Nalusa Philea, and it was described as being a dark, tall entity, one which inhabited forests. The creature would only come out when darkness encroached, and so appearing to shun the day and light. And as so many other similar stories of different cultures say, the creatures would then attempt to steal children.

2. Islamic Folklore: In Islamic folklore, there are mentions of the "Jinn" or "Djinn," supernatural beings created from smokeless fire.

Our modern idea of "genies" stems directly from the Islamic belief in such beings. Some interpretations suggest that these entities can take on shadowy forms and seem to inhabit desolate or dark places, such as abandoned buildings, old houses, or even modern apartments. And this is strange because many so-called "ghost hunters" have seen glimpses of shadow people in abandoned buildings and in some cases, have even captured video of them. So, the idea might not be as strange as it seems, that Shadow People inhabit dark and abandoned places, including buildings, tunnels, sewer systems (in larger cities), caves, etc. And people often report seeing Shadow People at night. While not exactly synonymous with the Western concept of Shadow People then, there are major similarities in the mysterious and sometimes malevolent nature of these entities. However, one should always avoid interactions with Shadow People, or "Jinn" as they are innately evil according to Islamic folklore. And, they may well be right!

3. Japanese Folklore:

In Japanese folklore, there are tales of creatures like the "Kage-onna" or "the shadow woman." These entities appear as human-shaped shadows and are associated with haunted places or curses. Again, we see here a similarity that seems a common threat with Shadow People, that they like to dwell or inhabit the environs of abandoned buildings or isolated places, such ones dark and gloomy by their very nature.

These "shadow women" may bring misfortune or be actual harbingers of tragedies to come, according to Japanese legends. In America, many people believe that the so-called "Mothman" is also a harbinger of such dangers. The Mothman is said to only appear at night or in dark places, has huge bat-like wings, glowing red eyes (the same as some Shadow People seem to have), and whose appearance is a forerunner to tragedies. Many claim that Mothman sightings occurred just before the Japanese Fukushima disaster, and actually have convincing photos of such sightings. Many people claim, as well, that the Mothman appeared at the New York's twin towers on September 11 of 2001, and The Silver Bridge Collapse at Point Pleasant, West Virginia, December 15, 1967. The "shadow woman" of Japanese legend is, apparently, not a human's fried.

4. African Folklore:

In certain African cultures, there are stories of shadowy spirits or entities that are manifestations of ancestral spirits or supernatural forces. These beings can be either protectors or warnings of something bad (again, harbingers of disaster), depending on the context of the visitation. In addition, as seems usually the case, such sightings get interpreted as usually being dangerous or evil in some way. Therefore, according to such folklore, they are best avoided.

5. European Folklore:

The very term "Shadow" People and the word "Shade" come from the same root origins. Shade comes to us from ancient Greek mythology They believed the ghosts traveled to an afterlife that they referred to as the "Underworld." In the underworld everyone there were but shadows or "shades" of what they had been in life.

In European folklore, there are various accounts of shadowy entities, often associated with the supernatural or the spirit world. These beings are often either omens or messengers, and their appearances often seem tied to specific events or locations. Again, the "Old Hag," is one example of this, or as is the "Hat Man" and various other named shadowy entities that are considered to be dangerous in themselves, or heralds of coming disaster. An interesting side note: our word, "nightmare" is derived from Old Norse. A "Mare" or more originally, a "Mara" was something dark that sat on you while you slept and which would then strangle or smother you. So, even our very word for bad dreams, "nightmare," seems to originate from the idea that something crawled onto people in the dark of night and strangled them! Whatever this phenomenon was that caused this, it must have been very common to result in our word for bad dreams being based on its occurrence!

6. South American Folklore:

In South American folklore, particularly in some indigenous tribal traditions, there are stories of shadowy entities or spirits, as well. These beings might be associated with the natural elements at times, and people there perceive them as powerful and mysterious creatures, often having an evil aura about them. They are to be feared and avoided, or at least, held in respectful awe.

7. Polynesian Folklore:

In Polynesian cultures, there are myths and legends that involve shadowy spirits or supernatural beings. These entities are protectors of certain places or seen as manifestations of the unseen forces that govern the world. Whether such forces are good or evil remains to be seen. However, according to Polynesian folklore it is best not to be singled out by such beings, because they are beyond our ken and therefore their motives can't be known but might just be malevolent. So, they are best avoided. This is not too dissimilar to the Nordic God, Loki, whose behavior often had unpleasant outcomes for humans. The malevolence and urges on Loki's part to manifest dark tricks on humans long ago earned him the nickname of "the trickster."

8. Chinese Folklore:

Chinese folklore includes tales of various supernatural beings, and some stories involve shadowy figures that may be associated with ghosts, spirits, or mystical forces. These entities might play a role in traditional ghost stories or tales of the afterlife. They are said to often interact with humans in a number of ways, such as appearing in dark places to them, manifesting themselves in their dreams, or even appearing in gloomy alleys and dark buildings. They are widely accepted as an integral part of Chinese folklore and many Chinese see them as being an entirely real phenomenon. In short, to many Chinese, they are not folklore at all, but a part of their everyday reality.

Again, it is important to note that while there are cultural similarities in the overall theme of shadowy entities, the specifics can vary widely. Different cultures have unique interpretations of such phenomena, often reflecting these on their beliefs, values, and understanding of the spirit world. However, the recurring belief that such entities are evil, dangerous, or harbingers of tragedies seems global and widespread. Even here, in America, almost all people interpret them as being scary beings and probably dangerous in some way. The feeling of unease and fear they engender seems to be a constant.

Chapter Conclusion: There are many commonalities in reports of Shadow People from around the world. Although interpretations of events regarding Shadow People may differ, the events themselves are, more often than not, very similar. Also, the idea of Shadow People being dangerous or malicious also seems to be a widespread one.

So how does one account for this, this worldwide so-called "mythology" of Shadow People? As with historical global stories of a Great Flood, which once were thought of as nothing more than "mere legends" by most archaeologists, but now are viewed by many historians as actual stories of real events, is there something more to the idea of Shadow People being real and not just mere legends? And remember, these legends are, literally, worldwide. Every continent but Antarctica has them, and that one exception is strictly because only a few researchers inhabit that frigid and desolate region.

Are such ancient stories of encounters with Shadow People, although changed and/or enhanced over the centuries, true? Do current global stories of Shadow People also have such kernels of truth to them today? It would seem that they just might. Whatever these entities are, they seem to have been with us for a long, long time! They seem to have always been here, lurking on the fringes of our reality, or if you will, in the shadows of our world. Additionally, whether rational or not, Shadow People have long history of being associated with dark forces and malevolence towards humans.

Chapter 5—Other Shadow People References, Including Religions

It is important to note that explicit historical records, specific references to Shadow People, as we commonly use the term today, are difficult to find. This is because the concept of shadowy or supernatural beings existing is so deeply embedded in cultural folklore, religious texts, and personal accounts that they are hard to extract as being in line with what we consider to be the definition of Shadow People today. As always when we discuss the history of this topic, it is subject to various interpretations based on who encountered such beings, when, and where. Over time, these reports can vary widely as to whether these entities are of a truly religious nature, such as visions of angels, etc., or simple entities of a paranormal origin, or indeed, even "something else" entirely. The following are examples of this:

1. Ancient Texts:

Various ancient religious texts and religious mythologies from different times, as well as from around the world, including those of Greece, Rome, Egypt, China, India, the Americas, etc., contain often cryptic references to entities or spirits that today we would interpret as shadowy beings (Shadow People). However, they were not considered to be of such a nature by those who recorded such things. For example, in ancient Egyptian "mythology," (that the perceived as their real religion and so true), including even in the Egyptian Book of the Dead, there are mentions of shadowy entities associated with the afterlife. These could appear in our own world at times and not just the Underworld. Were these Shadow People, or manifestations of the Egyptian gods, as those who believed this at the time might have thought? Alternatively, were they something else entirely?

Almost all religions, including both pagan and mainstream religions of today, and whether those of indigenous peoples, or mainstream western and eastern world societies, speak of such shadowy-like creatures appearing in various forms. Often these stories come with warnings to avoid such things whenever possible. For instance, the Catholic tradition of incubi and succubae, those male or female dark shadow beings that tempt people while they are in bed and/or on the verge of sleep, and which the Catholic Church had deemed to be evil demons are just one example. Were they demons, or are they just shadow beings as we think of them now, and not really demons in the biblical sense at all?

Or is there a difference if Shadow People really are malevolent, too, just as such so-called demons might be?

2. Medieval Accounts:

As we just saw with idea of the incubus or succubus, some medieval accounts and folktales of Europe and elsewhere during the so-called Dark Ages of history mention encounters with dark or shadowy figures, often seen in similar shapes and forms despite being spotted in widely disparate geographical locations. These and other such stories often blended elements of folklore, superstition, and religious beliefs. The result is that any sightings of Shadow People during those times are overlayed with dense embellishments of religious beliefs and/or superstitions. In short, because of this, it is hard to separate the "wheat from the chaff," as it were, with regard to such historical accounts.

Despite such "Shadow People" being attributed by people of those days to almost always being the result of the machinations of leprechauns, demons, fairy or fey folk, or whatever, again, the basic substance of the descriptions of such events are often amazingly similar to those of today. Moreover, this, despite the stories coming from different regions of the world and from different medieval sub-periods.

Always it is a creature of the dark, usually with only a vague form or even formless, that creeps forth from the shadows that they seem to inhabit. They make no sounds, utter no cries or noises, strange or otherwise, of any sort. In addition, far more often than not, there is a distinct fear element involved, as if such creatures are dangerous or at least, best avoided.

So, again, we have that fear. And why is that? Is it just humanity's fear of the unknown? We don't seem too afraid of angels, or fairies, for instance, so we don't fear all forms of supernatural occurrences. Nevertheless, with Shadow People, we do seem to harbor a fear of them and on a pretty consistent basis, not only around the world, but down through history, as well, it would seem.

3. Written Historical Accounts of Paranormal Phenomena:

In historical written records, one will sometimes even find accounts of paranormal phenomena or unexplained experiences that also involve encounters with mysterious "dark" or "shadowy" figures. However, again, the descriptions of these beings vary some in such writings, and the term "Shadow People" as we know it today, was not then in use. Nonetheless, again there is the commonality of the figures being "dark" somewhat vaguely "human-like" in shape, and/or just "shadows" without any detailed facial features and making no noises. In some cases, witnesses described them as shadows, ones that became distinct moving blobs ("amorphous shapes"). These are said to issue forth from dark corners in buildings and forests, and such places. Another commonality of such events.

These reports date back so far, that even the Ancient Sumerians, over 5,000 years in the past and one of the very first founding civilizations of humanity, talk of an Alu. The Alu was/is a tall creature of darkness that strode forth during the night. It would seek out sleeping people and seemingly drain them of energy/life to the point where many wouldn't wake up and instead would either fall into a coma or even die. Again, does this sound familiar? It should! It is an amazingly similar story to many other cultures and throughout history.

4. Diaries and Personal Accounts:

Diaries and personal accounts throughout history sometimes contain descriptions of unusual shadow experiences. Some individuals, even historical figures of some renown, have documented encounters with shadowy figures or entities, attributing them to supernatural or inexplicable causes. Most often, they relate them somehow to their then-prevailing religious beliefs. Often, they saw them as signs and omens of something awful soon to happen, and so appear as warnings to such individuals. This is a recurring theme in such accounts and is a commonality that transcends them all. It might also be a reason why they are so often feared.

This is important and cannot be overstressed; usually such sightings are seen through the filter of the individual person's belief systems they hold at the time of the sighting. This means that often, some attributes assigned to such Shadow People sightings may not be accurate. The sightings themselves seem real enough, but the person who witnesses such an event often thinks of them in terms of their then currently held religious beliefs. Such events are then filtered through the perspectives of the viewer of the times and often altered to fit their current belief systems.

For example, although rarely do such sightings involve anything overtly or intrinsically "evil" or overtly dangerous about them, many people, the majority in fact, and I include myself in this group, subjectively view them as being evil in some way. Furthermore, some people interpret such sightings and their consequences as signs of the devil, demons, etc. Others seem them as messengers of a sort.

Why is that? The answer to that question is that we simply do not know...yet. In addition, a part of me wonders if we even want to know. After all, "where ignorance is bliss, a little knowledge is a dangerous thing," as the quote by Alexander Pope goes. I wonder if this the case here? Should we leave well enough alone, perhaps? Or does danger lay in that direction, as well. What we don't know could just as easily hurt us as much as something we do know!

5. Occult and Esoteric Literature:

It is isn't just encounters of the times there were deemed of as being of a religious nature. Other writings speak of Shadow People in ancient times. Some of the more occult and esoteric literature/manuscripts/books from not only prior to and during the late medieval period, but even after that period, contain references to shadowy or spectral beings that seem very similar to the descriptions of Shadow People of today. These writings often delve into mystical and paranormal phenomena, as well, such as the journals and diaries of some alchemists. Some such writings had been banned by the Catholic Church of the period and had been declared heretical. Often, these books were kept hidden from church officials as a result.

Today, even people who take certain mind-altering drugs, such as psychedelic drugs, both naturally occurring ones and some that are manmade, repeatedly speak of different users seeing beings, such as "machine elves" and other, sometimes "shadowy" entities, too. So, it isn't just religious texts or folktales that speak of such entities.

Even today, people speak of such entities as being real. With regard to those who take drugs and see such beings--are these just hallucinations on their part, or might there be something more to them? Why is it so many psychedelic drug users, especially those that take ayahuasca see the same thing? Could it be such drugs alter our physical perceptions just enough perhaps to give us a glimpse at the true nature of reality in its entirety, or again, are they just drug-induced hallucinations after all and so just a weird coincidence?

Moreover, if this last is so, why do those who take such drugs so often all consistently see the same sorts of hallucinations, as with the so-called "machine elves" for those people who partake of the drug ayahuasca?

Chapter Conclusion. As mentioned in this chapter, traditional folklore, cultural myths, histories, and legends also speak of Shadow People and often. However, remember, it is crucial to approach historical records and such folklore with a degree of skepticism, and always with considering the cultural and contextual factors that influence interpretations of such events. Moreover, religion has a strong impact on how people view such sightings, and so how they interpret them as a result, whether as angels or demons, for instance. Is it all just in the eyes of the beholder, then, as to how we interpret such sightings, or is there more to it than that?

For instance, a modern pagan may not view the sighting of Shadow People the same as a modern, Born-Again Christian might. A pagan might see such a sighting as a manifestation of the fairy folk, elves, leprechauns, or gnomes, etc. Whereas, a Born-Again Christian would probably interpret such an event as the work of Satan or a demon. For that matter, a scientist might not view such a sighting either way, but instead take a more pragmatic and perhaps scientifically realistic approach, even to the point of applying the Scientific Method of discovery to learn more about the matter.

This last raises an interesting point; does the Scientific Method work for all things, or is the realm of the metaphysical beyond its boundaries? Does the Scientific Method only work to the reality that we can perceive around us, or can it extend to include a possibly greater reality, one even of other dimensions, perhaps? The jury still seems to be out on the answer to those questions.

One thing, however, is certain; the Scientific Method does seem to have limitations when it comes to things of a seemingly metaphysical nature. For one thing, the method relies heavily on being able to allow others to repeat an experiment to prove its validity. How does one repeat such experiments when it comes to the subject of sporadic ghostly visitations or brief appearances of Shadow People, if they truly exist? What if they are a "one-off" phenomenon, never to be repeated at a given location? How then does the Scientific Method apply, or can it? Without repeatability of the event or some other independent corroboration of it, the Scientific Method doesn't work very well at all.

However, there are some researchers attempting to do this very thing, attempting to gather scientific proof of such possible entities, and some of those even use electronic instrumentation in this endeavor. The results of such attempts are tantalizing, but they are not seen as conclusive by most researchers. Yet, despite this lack of a "smoking gun" when it comes to proof of ghosts or Shadow People, the Scientific Method is the only real means, the only tool we have of approaching the matter in such a way as to get concrete evidence of such beings. So, we have to attempt to use it, at the very least.

Why is this important, using the Scientific Method? Because, as just mentioned in the prior paragraph, if we want to arrive at the truth, however good or bad for us it may turn out to be, we have to take the scientist's approach, the realistic one whenever possible. Without the Scientific Method, it is very difficult to approach matters of the paranormal in any meaningful way. The same phenomena can result in very different explanations and interpretations for them by different people otherwise, and that's with no reliable way to determine which explanation is the correct one.

For instance, the very term "Shadow People," itself, is a relatively modern concept used to describe a specific type of paranormal encounter. But this wasn't always so. As an example of this with regard to UFOs, an historian who traveled with Alexander the Great's army spied silvery flying "shields" above them while near the Indus River during Alexander's march into India. These things flew overhead, and were seen literally as "silvery shields" as they were prone to describing UFOs at the time. Since such sightings were out of the ordinary, they related them to something that already existed. Beings silvery in nature, and roughly the shape of shields they thus became "flying shields." Also, because of their highly unusual occurrence, they, they flying shields were viewed by the people of the day as signs from the gods or demons.

Today, we would perceive those "shields" as being flying saucers or UFOS/UAPS and viewed by most people as having nothing to do with either God or Satan. Nevertheless, what we need is *to actually know*, not just what we *"think" we know* these objects are. Obviously, they are not, in reality, either "silvery shields" or "flying saucers." Therefore, whether shields as ancient folk see them or flying saucers as modern folk would see then, the sightings are still essentially the same, just named differently, and so perhaps perceived differently. *But just what are they really?*

Chapter Conclusion. What is their real nature of Shadow People? That is still unknown, but it is imperative for us to find out the answer to that. After all, as with UFOS, that invade our airspaces seemingly at will and defying the conventional laws of physics as we know them just might be dangerous! It is also just as imperative to know what Shadow People really are. Are they even real? If so, are they dangerous or a threat to us in some way? Where do they come from? What are they? Have they always been here, or are they something more recent? These are questions we need answered and the sooner the better, if only for our safety.

Chapter 6 — Commonalities in Accounts of Shadow People

It is important to note that many reports of Shadow People from around the world today, do share some definite commonalities, just as they did in the past. Moreover, just as in the past, there are also notable differences in descriptions due (as I believe them to be) being based on the available evidence but largely also due to local cultural interpretations of such events. The sightings are thus altered based on such interpretations. Where an Islamic person today might see such a sighting as evidence of a djinn or genie, a Christian might see them as either devils, demons, or even angels. However, despite such different interpretations of these sightings let us focus now on the commonalties of such events.

Commonalities:

1. Shadowy Figures. Today, reports of such entities come from around the world, and across very different cultures. Even so, there is a consistent theme of Shadow

People appearing as dark, shadowy figures, but often with no further description or details in most such cases. These entities, often described as having humanoid shapes, always seem to lack distinct facial features or detailed characteristics, except perhaps, of the most basic sort, such as glowing red eyes, and that last, only sometimes. Shadow People even appear as just dark shapeless blobs emerging from shadowed corners of rooms or even outdoors from pits of darkness in streets, alleys, etc. They can be seen in people's homes. Always the sightings, whether indoors or outside, are fleeting in appearance. Shadow People can be seen to wear outlines and shapes that resemble clothing or apparel of some sort, such as coats and hats, as well, but all is composed of shadow, so such sightings are without specific detail or any color to them other than being described as "dark" "gray," or "black."

The Hat Man?

2. Nighttime Encounters. Many, perhaps most reports, involve sightings of these entities as being during the night or in low-light conditions. The darkness seems to enhance the visibility of these shadowy entities, leading to increased reports of them as occurring mostly during nighttime hours. It may also be that whatever these things are; they prefer dim light for whatever reason(s). They might simply be nocturnal, that is creatures of the night, as well. Nevertheless, they do seem to shun bright lit situations in the main. Shadow People can be seen in the daytime, as well, but such sightings are often extremely brief. And with regard to nighttime encounters, there was the case reported by the Los Angelese Times regarding some people from Laos (Laotians) who came to America as a result of the Viet Nam and Laotian Wars.

American intelligent agencies recruited a local people in Laos to aid them in their war efforts. These were called the Mong. Many of the Mong ultimately ended up in the United States. However, a strange thing then occurred; perfectly normal Mong men, mostly young ones began to not survive the night. They would die, seeming without reason, while asleep.

At first, doctors weren't too concerned, but as the numbers of young men dying topped 100, their concerns became acute. No cause, however, was ever found for the deaths, and they did stop shortly later. What do the Mong, themselves, believe as to what happened and why? They believed it was a thing of darkness, a creature of the night (supernatural in nature) that would climb on to the young men's chests and stay there till the men died of suffocation or being smothered. Their term for this creature was a "pressure demon." And that explanation is as good as any, because to this very day, doctors still have no clue as to what caused all those young men to die in their sleep!

3. Peripheral Vision. Witnesses most often describe seeing Shadow People in their peripheral vision rather than when looking directly at them. This is mostly true; however, there are some exceptions. Some people not only witness them directly in sight, but can and have managed to sometimes capture them on camera. In any case, this phenomenon of seeing them only briefly and/or peripherally contributes to the general idea that these entities are elusive, perhaps indefinable, and difficult to observe directly.

It may be that viewing Shadow People directly is somehow difficult for any of a number of different reasons. For instance, whether this is an inability to just be able to see them "face on" or if it's just "the nature of the beast," or something else, perhaps due to the limitations of some sort of cloaking ability on the part of Shadow People, or whatever, again, most sightings are via a person's peripheral vision.

1. Feelings of Fear.

Again, this is one aspect of Shadow People that I have personally experienced, an unreasoning (or maybe it is not so unreasoning) fear at such sightings. Nevertheless, encounters with Shadow People are more frequently than not associated with inexplicable feelings of fear, terror, or other distinct sensations of discomfort and unease. Regardless of cultural background, individuals often report an unaccountable sense of dread or foreboding during these experiences, and this, despite the Shadow People not seeming to pose direct or overt threats to the viewers during such appearances.

5. Brief Appearances.

As mentioned, Shadow People sightings are typically very brief and fleeting. Again, there are some exceptions to this, but even the longer encounters are no more than mere seconds in length, or perhaps just a minute at most. These latter viewings are extremely rare, though. When it comes to seeing Shadow People, brevity of appearance seems the rule rather than the exception.

Shadow People may disappear or move quickly, very quickly when noticed, adding to the mysterious nature of the encounters. Moreover, material objects don't seem to be obstacles for them. They appear to have the ability to pass through walls, for instance. A Shadow Person can be seen more than once in a short period. However, always the appearances seem very brief, even so.

6. Shadow People don't make noise.

Sightings of Shadow People always seem to be silent. That is in the vast majority of cases, but not quite all. They appear then disappear, but usually make no noise. No footsteps are heard, no cries or screams made and again, this is with very few exceptions. They are truly just as shadows, bereft of sounds and voice, and left only as dark shapes of an unnatural nature. Or are they?

Chapter Conclusion. Now that we have an idea of just how Shadow People appear in today's world, the commonalities of such sightings, let's delve into the differences observed in such sightings. We will do this in the next chapter.

Chapter 7—Some Differences Reported

In the last chapter, we cited some of the similarities and commonalities reported in Shadow People sightings from around the globe in today's world. Nevertheless, there are some differences? These, in the main, are:

Cultural Interpretations: Yes, we have been here before but this is a crucial aspect of Shadow People and, therefore, deserves a bit more in-depth explanation. The cultural interpretation of Shadow People varies widely depending on the different locations in the world where such sightings occur. As mentioned earlier, this was also true in the past, but is just as true in today's world. Just as people often had their own interpretations in the past of such sightings, so, too, it is the same in the present. Some cultures of today view these entities as protectors or messengers, while in other cultures; they are associated with evil, malevolence, and even as harbingers of dire things to come and so are seen as signs and portents or fearful omens. Again, this is much like the range of interpretations from more ancient sightings. Some things never change, it seems, or at least, people don't seem to.

Spiritual or Supernatural Associations. Different cultures also attribute Shadow People to various spiritual or supernatural origins. This is true in many cases, including right here in the United States. Some see them as ghosts, spirits, demons, angels, or even interdimensional and/or extraterrestrial entities. There are also individuals who believe that Shadow People are spiritual entities of some unknown sort. Some think they might be extraterrestrial in nature. As always, depending on the cultural, religious, or personal beliefs of the individuals involved, these entities even can be seen as omens, or as messengers conveying communications from the spirit realm. Amazingly, the idea of just what a Shadow Person is can vary widely from individual to individual, even when both are products of the same culture. This gives one an idea of just how at sea we are when it comes to explaining this strange phenomenon. We just don't know much about them, and so every type of explanation conceivable comes into play.

Specific Geographic Characteristics.

Geographic/Cultural differences, just as in the past, also influence the specific characteristics attributed to Shadow People seen in differing geographical areas today.

For example, in Japan, in Japanese folklore, there are tales of the "Kage-onna" or "shadow woman," and often current sightings are ascribed to being manifestations of her, even now. Whereas in Western cultures, the figures may be more generic in appearance. However, there are exceptions to that idea, as with the "Old Hag," appearances in Europe, as well as those of the "Hat Man," and certain others here in the Americas. None of the examples mentioned in this section are of what could be called positive visitations, since people are warned about these same apparitions as being harmful or dangerous and possibly even being life threatening. Which leads us to the next point:

Omens or Harbingers of Evil? In certain cultures around the world, but certainly not all, Shadow People are seen as omens or harbingers of future events that are not going to be good ones in their nature. Even as in the ancient western world, where comets were seen as ill omens for centuries, the same holds true for various cultures with how they view Shadow People sightings today.

In other words, if you believe in such entities as beings that can foreshadow things to come, then those things to come are not going to be good! It is important to remember here that the depth and degree of the nature of these events and the cultural significances attached to them can vary significantly from culture to culture, geographically, and again depending on the local beliefs of that particular region. We do have to ask, though, why they are ill-omen harbingers and not positive ones...ever, and without exception?

Chapter Conclusion. One thing that cannot be overstated here; although Shadow People seem to appear in some form or another worldwide and throughout the ages, the details of how they exactly appear, what they mean by such manifestations, and why they appear as they do, can vary significantly from place to place and time to time, as well as from culture to culture. This can make it hard to sift through all the legends and myths, ancient and modern, to ascertain the real facts concerning such sightings. However, we need to do this. We need to know just what it is all about, if only for our possible future safety as individuals and perhaps as a species.

Chapter 8—Types of Shadow People

Shadow Person on Stairs?

This is just a quick list of the different types of Shadow People that seem to appear. Although perhaps not all types of them are listed here, the main ones are:

Appearances:

1. Shadowy, human-shaped creatures, sometimes seeming to wear coats and/or hats. Always appear black or dark gray with no other distinguishing details.

2. Shadowy creatures that appear more as large, upright blobs, but without any distinctive features at all. Always these appear as black or dark gray.

3. Shadows only, appearing two dimensional as if real shadows of something, something that isn't there to cast such a shadow They vaguely resemble a human shape but are incredibly swift in movement and disappear abruptly.

4. Tall, amorphous dark figures with red eyes. Other than eyes, they appear black or dark gray and without any other distinguishing features. Some describe them as moving blobs.

Abilities and limitations of Shadow People:

1. Such creatures do not seem to have physical reality in the sense they can be felt or even feel anything else. There is a caveat to this, though, as with the Incubus and Succubus that seem to be able to physically sexually assault humans, and also there are reported incidents of things/objects being knocked over or "left out of place.".

2. They do not seem to be able to move objects as a rule, and never seem to carry anything or possess anything such as a weapon or object of any sort. However, again, some objects near them sometimes can fall over, but this seems rare.

3. They do seem to invoke a strong sense of fear, but whether this is an ability on their part, or just the natural reaction of people to seeing something so bizarre is uncertain.

4. Again, they can move extremely quickly and disappear just as quickly as they appear. In some cases, it appears they can move out of locked rooms and so can pass through walls.

5. They never speak or seem to make sounds. However, often, they can be seen leaning out from behind some object, as if spying on people.

How they interact with humans:

1. Shadow people seemingly can appear anywhere, whether indoors or outdoors. They have been seen on streets, in yards, the woods, but also indoors in rooms, basements, attics, kitchens, etc. Walls and locks don't seem to stop them.

2. As a general rule, they don't seem to be physically threatening, and flee or simply vanish when chased or approached. However, again, they do seem to always strike fear to some degree in the viewer. This seems a very visceral sort of fear.

3. Shadow people more often appear at night, in dark areas, but sometimes in the daytime (usually indoors on such occasions). Outdoors, it is usually at dusk or later in the evening. It is as if they shun light.

4. On occasion, Shadow People can seem extraordinarily sinister or dangerous in a real sense, not imaginary. There is one video on the Internet showing a shadow creature, dark arm outstretched and moving inward between the slats of a baby's crib. A nanny cam picked up the incident during the night while monitoring the baby. When startled, the shadow hand and strange body shape attached to it withdrew, but honestly after seeing this video, that baby, crib and all, would be moved out of that room right away if I were the parent involved! The video was judged not to have been a hoax and seems all the more frightening as a result. Some viewers of the video even wondered is such a thing might be what causes SIDS (Sudden Infant Death Syndrome) in some infants. Whether there is any merit to such a hypothesis not known at this time. More information is needed to ascertain if such an idea might be valid.

Chapter 9—The Psychology of Shadows

Now we come to the psychological aspects of Shadow People visitations and their seeming reason for such intrusions upon unsuspecting humans. First, what are some of the possible explanations for Shadow People sightings? Well as mentioned earlier, there are a number of mundane explanations for many such sightings, but these explanations do not seem to apply to hardly all such visitations.

1. Sleep Paralysis Hypothesis. When people fall asleep, their bodies shut down most forms of physical movement. It is conjectured this is to stop sleepwalking, or other types of movement that might be hazardous to an individual who is asleep and so unconscious. During the onset of sleep paralysis, people often enter a stage, where if they sudden become conscious again while still in the grip of sleep paralysis, they can often feel a strong sense of fear and sometimes can have strange delusions or hallucinations of some presence being near them. This then, might account for sightings of some Shadow People.

Problems with Sleep Paralysis. Although people who sometimes think they see Shadow People when they are awakened from their sleep during the night, sleep paralysis does not account for those who are fully awake, active, and either indoors or outdoors when they experience such a sighting. Nor does such an explanation account for photographs and videos taken of such strange anomalies. If it is captured on camera, it's probably not someone's imagination or just the result of sleep paralysis.

Sleep Deprivation. Sleep deprivation, going long hours and for days, can result in hallucinations and often does. These hallucinations can then account for sightings of Shadow People. However, again, the vast majority of sightings are not made by sleep deprived individuals, so this explanation, too, doesn't seem to work well for Shadow People sightings.

Substance Abuse. The taking of different types of drugs may cause some people to have visions of Shadow People. As mentioned earlier, ayahuasca users often talk of seeing "machine elves," or "machine people," as they call them. Nevertheless, the vast majority of Shadow sightings aren't made by users of this particular drug, or mushrooms, LSD, or other such mind-altering substances. So again, as a blanket explanation for Shadow People, substance abuse doesn't seem to work. In fact, some argue that ayahuasca actually might facilitate the ability of humans to see Shadow People.

Influence of stress, anxiety, and mental health on experiences. People suffering from certain mental health disorders, and/or stress and anxiety might be more prone to seeing visions of Shadow People but as yet, there has no meaningful correlation made by health experts with regard to this. They see that people with such mental problems are only slightly more prone to see Shadow People than people without such problems and so little is the difference between the two groups, that statistically there is no real difference. It should be said here, though, that health experts think the matter should be delved into more deeply, since no real studies have been made on the subject to date.

Migraines. Some psychologists and psychiatrists wonder if migraines might also be a cause of seeing Shadow People. Migraines, those intensely painful headaches that can people to have tunnel vision, or see auras, or even shun light, might be a cause of seeing Shadow People through hallucinations. However, there is little evidence to support this theory and a lot against the idea, frankly/

Collective Unconscious. Some psychologists propose the idea that Shadow People like the Hat Man are manifestations of the "collective subconscious" as described originally by the famous psychiatrist, Carl Jung, a contemporary of the so-called father of psychology, Sigmund Freud. Whether this is true or not, remains to be seen. One thing is certain, since Shadow People can sometimes physically impact our reality, they would have to be a very powerful manifestation of humanity's collective subconscious, indeed!

Shadow Self. Some in the psychiatric field wonder if Shadow People aren't really the "Shadow Self?" This is an idea also promulgated by Carl Jung in which he believed regular people might be able to throw a sort of projection of themselves as a "Shadow Self." These Shadow Selves could have some power over their physical environments being a manifestation, as it were, of a living human.

Chapter 10—Scientific theories about Shadow People

Here we come to a rather tricky area on the subject of Shadow People. Most scientists dismiss the idea of Shadow People outright for lack of any evidence, but some researchers have come up with some interesting ideas for just what these things might be:

1. Quantum physics. Some researchers with backgrounds in quantum physics wonder if Shadow People might belong to a different "realm" of reality. In other words, they are part of a greater reality than we can sense. This approach seems to have some merit. After all, the human eye can't see the entire spectrum of light. Our vision stops at infrared on the low end of the spectrum and ultraviolet at the top. Some people can't even see the color indigo because it is too close to the end of visible spectrum for them. So, the idea these creatures might be a natural part of the universe and of Earth cannot be ruled out. This crosses over into metaphysics at some point.

2. Metaphysical Interpretations. In the absence of any other explanations, some have turned to the metaphysical for answers. Here, various researchers have entered the realm of the unanswerable in that it may be impossible to prove these things are of a metaphysical nature or even if they exist in the normal sense of the word. Even though they could well be the product of such a thing. Still, by its very nature, the metaphysical interpretation may be a dead end for researchers since such anomalies more often than not, defy the attempts made to apply the scientific method to them.

Parapsychological investigations and research findings. There are bona fide amateur and professional investigators of Shadow People. Many times, fleeting views of them have been caught on camera, as a result. The Internet, specifically YouTube, but also on other media social groups, videos are rife showing glimpses of these strange beings. **Attached at the end of this book, please find links to such videos on YouTube.** I think you will many of them interesting.

Are some such videos hoaxes? Oh, without a doubt; one can't escape faked photographs and videos these days, but some are well credentialed, and have been proven not to be fakes, as well.

Anecdotal evidence vs. empirical data. There is far more anecdotal (stories) about sightings of Shadow People, than actual hard evidence. However, this has only been true until recently. With the advent of the common usage of phone cameras, etc., that is now changing. Added to this, the number of ghost hunters and researchers in the paranormal have exploded in recent years, and again, they are accumulating real evidence in some instances. Just check out some of those links I've provided at the end of this book for verification on this fact. So, the times are changing and at last, we may just begin to build a real and solid data base of hard evidence concerning Shadow People. In fact, this is already happening.

Chapter 11—Shadow People in Pop Culture

Shadow Person Again?

Now we come to the subject of Shadow People and how they are impacting us today in our modern culture. And to say they are, is an understatement! Shadow People have been making headway as a subject throughout social media, in literature, very much in films and television to a slightly lesser extent.

Is this important? Is it relevant to us whether Shadow People are a making such inroads into our pop culture? Well, yes, it is, because such references, depictions, and tales about Shadow People color our perceptions of this mystery, and they do it just as much now, as perceptions of individuals in past ages colored their perceptions.

Shadow People are almost always depicted as scary beings, and more often than not having evil intentions. The anecdotes about sightings of the "Hat Man," or stories of the "Slender Man," or even variations on the "Old Hag," and even "The Grudge" are examples of this. Even movies, like "The Ring" are just offshoots of the "Old Hag," idea, where a younger woman is used, but with all the attributes of the historical "Old Hag."

A.I. Image of Old Hag.

The "Hat Man" has made real inroads into our culture in this regard. He appears remarkably similar in all accounts of visitations by him, no matter where and when these occur. Being tall, with a slender shape and seeming to wear a hat of some sort, one with a large brim, this being can appear in darkened rooms, usually bedrooms and often as the person is just falling asleep, but often also when they are wide awake.

The descriptions of this shadowy figure are amazingly similar and multiple reports of visitations by him, from people in widely different geographical areas, are compelling

In fact, the "Hat Man" has made such inroads into our culture that even a series of hit movies were based on him. Ever watched any of the Freddy Krueger, Nightmare on Elm Street movies? It has been said that Wes Craven drew part of his inspiration for the character from the "Hat Man." The similarities to Freddy and the Hat Man are striking. A shadowy figure with a brimmed hat appearing as if a "nightmare" to people, and just as the Hat Man is said to do, materializing in people's bedrooms just when they are almost asleep. So there has definitely been representations repeatedly of Shadow People in some form or fashion in movies, television shows and certainly in literature, as well.

With regard to the Hat Man, some people claim to have also seen him while using the drug, DMT. This is a strong, mind-altering drug that causes hallucinations and/or visions. It is considered to be a psychedelic. It can be produced in our pineal glands at times and some researchers think it might be exuded by the gland when people are suffering near-death experiences. in the brain, and some people believe that it is released during near-death experiences, dreams, or meditation. People also use DMT with ayahuasca, this is a brew that was and is used in the Amazon, but it is not the same thing as DMT, although DMT is an essential ingredient in ayahuasca. The Hat Man is being seen sometimes by certain users of DMT. They are various reports about how the Hat Man users report seeing the Hat Man during their trips He can behave different towards different users of DMT, being kind and helpful to some, but a torturer of others. Furthermore, many of these users are convinced the Hat Man is not a hallucination or figment of their imagination, but is a real being, one who just exists in a different dimension or real of existence.

Influence on Urban Legends and Contemporary Folklore. One can readily see simply by checking out social media, movies and literature that Shadow People have made a marked impression on the rest of humanity. This impression hasn't been generally good in most cases.

Such impressions have only served to reinforce the idea that shadow people are something to be frightened of…even when they may not be at all real. The "Slender Man" is a good example of this. He was a fictitious Shadow person and strongly resembled the Hat Man, except without the hat, having a very slim build. He impacted a lot of young people on social. Media and very negatively, as well, to the point of Murder! As Wikipedia put it:

*"On May 31, 2014, in Waukesha, Wisconsin, United States, two 12-year-old girls, Anissa Weier and Morgan Geyser, lured their friend Payton Leutner into a wooded area of a local park and stabbed her 19 times **to appease the fictional character Slender Man.**[2] Weier and Geyser were both found not guilty by mental disease or defect and committed to mental health institutions. Weier received a sentence of 25 years to life and Geyser was sentenced to 40 years to life. After seven years in custody, Weier was granted early release and will be under supervision until age 37."*
[Emphasis added.]

So, pop culture has definitely been impacted and quite strongly by the concept of Shadow People. It should be said, that based on the horrible account with regard to children believing in such things as the "Slender Man," some attention should be given to the inherent dangers of this with regard to children and even some adults who may be too easily swayed by such urban legends as that of the Slender Man.

Chapter Conclusion: Do Shadow People make a big impact on our current culture/pop culture. Most assuredly, they do. This is partly good and partly bad. It's great that the idea of the existence of Shadow People is becoming more accepted, because this sparks people's interest in the subject, want to learn more about it, and encourage researchers to take the idea seriously and to investigate the matter more thoroughly

The bad side of this rise of Shadow People in pop culture is that it also gives rise to urban myths and conspiracy theories, some of which then evolve into dangerous variations, as well as sometimes ones with deadly consequences as with the two murdering girls who did it for the sake of the imaginary "Slender Man."

Chapter 12—Encounters with Shadow People

Photo of Shadow Person

First-hand accounts from individuals who have encountered Shadow People are numerous, for specific individual examples, please follow the links at the end of this book to sites where such videos and commentaries are located. However, these below, broadly, are the types of encounters most individuals have had:

1. Most individuals just cast glimpses of Shadow People, short-lived, and often too quick to get evidence of with a camera.

2. Others see more persistent visions of Shadow People, usually when they are in bed and spy a dark shape standing in their darkened bedroom. The "Hat Man" is a good example of this.

3. Of late, a lot of ghost hunters are catching sightings of what appear to be Shadow People. Fleeting images of shadows moving across a room, an adjacent corridor, or leaning out from the entrance to another room to regard the researchers. When chased to dead ends in the buildings, these shadows simply disappear. Researchers find such interactions always frightening and upsetting.

4. Some Shadow People appear in rooms that are being monitored. These Shadow People seem to materialize in children's rooms, even forming below their cribs and then reaching up toward the babies, all while being caught on monitoring cameras. Shadow People also appear in adult and juvenile rooms that are being monitored via nanny cams, etc.

5. Some Shadow People manifest in dark alleys, streets late at night, or abandoned buildings. They can also appear after hours in office buildings, department stores, warehouses and more, often to the detriment of some unhappy security guard who must investigate such things, and usually all by him, or herself while doing so.

6. Shadow People have even appeared in isolated and rural roads, appearing on empty stretches of dirt lanes and even highways. They can suddenly manifest even in the woods in some places.

With regard to this last; there is a compelling encounter that occurred in Mexico or perhaps even a little further south than that country. I'm unsure of the exact location. This is actually a video and can be found on the Slapped Ham paranormal website. Several people are sitting on a porch of a house situated on the edge of a dirt road in a very rural area. It is dusk. The people are laughing and joking and seeming to have a good time. Only one man, on the far left of the group, seems to be looking out toward the camera, which appears to be situated across the road from the house. It seems someone had simply placed it there to record the get-together on the porch and then had left the camera on a tripod unattended.

A car, clearly visible drives past the house and people sitting there on the veranda. Then almost immediately after, a blurred shape, which might resemble someone on a motorcycle or might not, shoots down the road after the car. There is no noise from the thing, whatever it is and it is about 95 percent invisible. Nobody on the porch even notices it. They just carry on as if nothing had happened. There seemed to even be some dust raised from the passing "whatever it was."

However, the man at the end of the porch did see something. In the video, he immediately rises to his feet and turn his head to look down the road. Then, obviously still bewildered, he steps down from the veranda and walks into the middle of the dirt lane and stares after whatever it was that had just passed, obviously trying to ascertain what it was.

Invisible motorcycle or Shadow Person on such an invisible vehicle? "Something" clearly raced by, but just what that "something" was remains to be determined. So, are Shadow People real and just "cloaked?" do they choose darkness because it helps their cloaking devices to perform better? Is this why they seem to avoid bright lit, or daylit areas? The answer to that is still uncertain, but perhaps it could be the reason why these beings seem to avoid light.

With regard to cloaking devices, is this even possible? Well, yes, it is. Many countries are working on such devices and are achieving good results. Different approaches to the problem of invisibility are possible and are being tried. For some researcher, the use of lenses and more importantly, nanotechnology are the way to go to solve this problem. As a result, a number of nations have come close or already have manage invisibility for their planes, ships, and even submarines. Although referred to as "stealth" vehicles, because they have not as yet arrived at total invisibility, there are many countries, including the USA and others that are accomplishing something that was once thought of as being totally science fiction. As far backs as 2018, we discovered a substance now named metalens that made test devices undetectable, but this was just across the visible spectrum. A Japanese scientist, Professor Masahiko Inami, invented a true invisibility cloak that used camouflaging technology and it has had brilliant results. So why couldn't Shadow People have already achieved the same results?

Chapter Conclusion: Shadow People seem to appear everywhere and under multiple circumstances, but almost always it seems they appear at night, dusk, or dimly lit places. Are they actual beings that are simply invisible to us, or "cloaked?" They could well be, since they do seem to be able to disappear at will. Perhaps, then, they either naturally can appear cloaked to us somehow, and only can be seen under certain circumstances?

Chapter 13 What Are Shadow People?

Photo of a Shadow Person?

Now we come to the crux of the thing; just what are Shadow People? As we have mentioned, they have been described as supernatural beings, genies or djinn, spirits, ghosts, etc. Some even say they are just echoes, or reflections of people that once existed.

Trying to prove or disprove such a thing is a monumental task, although some researchers, ghost hunters, and others are attempting to apply some of the Scientific Method to at least get some answers in this regard. So far, despite making some progress in this regard, they still seem to have a long way to go.

What if Shadow People aren't spirits or ghosts? What if they are real entities and just a part of reality we can't normally see or feel? I can't see infrared, but that doesn't mean it doesn't exist. I can't see or feel dust mites but they exist by the thousands in our beds, carpets, etc. We can't see tardigrades because they are too small to see with the naked eye, but they, too, exist.

Alternatively, could Shadow People be a part of a greater reality that we have yet to discover?" Are they like the infrared or ultra violet of the visual spectrum to us, there but invisible to us? Or are they deliberately cloaked beings, hiding their existence from us? Do they exist in our world alongside of us, but hide this fact from us for the most part? The jury is still out on that one.

Let's look at this objectively:

1. We, ourselves, are developing closing devices, so what if "they" already have, and perhaps a long time ago? If we can manage it with out current technology, why couldn't some other, perhaps more advance people do the same? It's eminently possible.

2. When Shadow People do appear, they seem to take forms that are seen often and consistently, whether is just "dark amorphous blobs" or having some sort of vaguely humano9id shape. Would our imaginations always repeat this false vision? Or is there something more to Shadow People than just our imaginations.

3. Evidence for Shadow People seems to be real. Not only do reports of them date back centuries, but possibly millennia. Videos, photographs, and first-person testimonies by the thousands would seem to support this statement.

4. Shadow People sometimes, if only rarely seem to have a physical impact on our world/reality. Despite just being "shadows," Shadow People seem to be able to move objects at times, knocking things over, and even tossing things at people sometimes. As in the case of the people on the porch video, there is even evidence they might be using cloaked vehicles. So, can they still simply pass through solid walls? Or instead, are they turning their cloaking devices up to "Maximum" and thus seem to just totally disappear?

Now, some of the above information has been repeated here in this chapter for a reason. If we are to figure out just what Shadow People are and why they are, we have to know all their characteristics, or at least, as many as we can. Similarities of encounters with them can go a long way to helping us decide just what they are or are not, and also what they be capable of doing.

Standing Waves. Some researchers wonder if Shadow People and/or ghosts might not be "standing waves." A standing wave is where the energy can pass as waves through something, like water, for instance. Although the energy passes on, the wave in the water stays in the same position. This phenomenon can be seen from airlines when looking down on the ocean. The waves appear as if "frozen." They are not. The definition of a Standing Wave, according to Wikipedia, is…

"a wave that oscillates in time but does not move in space. It is also known as a stationary wave…Formation…Standing waves are created when two waves traveling in opposite directions interfere with each other." **[Emphasis added.]**

So, in the case of Shadow People, they could be a form of energy that maintains its shape and structure by being a sort of "psychic" standing wave.

Conclusions Based on the Evidence so Far:

Based on the above-described attributes listed in the previous chapters, I think the evidence shows (based on a preponderance of anecdotes, photographs, and videos, as well as centuries' long history of sightings and interactions with Shadow People) that Shadow People are:

1. Most definitely real. They seem to be real manifestations of "something." They appear far too often and under similar circumstances to be discounted as illusions, delusions, or hallucinations. How can one take videos and photos of hallucinations? One can't, so what is videoed or photographed, at least in some cases and discounting hoaxes, so they must be real.

2. Physical Impactive. They can physically impact this world. If they can use cloaked vehicles, knock over brooms, move tables, and such, then they can have a direct and physical impact on our world and us.

3. Cloaking Devices. They may well be using cloaking devices. In this way they can slip in and out of buildings, disappear at will, and learn a great deal about us. They can also use such devices to get into our homes, interact with our children, pets, and/or us. They do not appear to helpless disembodied beings.

4. **Knowledge.** They seem to have abilities beyond are present knowledge of physics or science, at least to some degree. If they have such marvelous cloaking devices or alternatively, can really pass through solid objects such as walls, then their knowledge is at least as extensive as our own, and probably more so. Also, if they are capable of this then they are:

5. **Sentient Beings.** One can't accumulate such knowledge, have such capabilities and not be considered conscious intelligent (sentient) beings.

6. **A Danger?** They may or may not present a danger to us, since they have all these capabilities, and it is the red-eyed Shadow People that are usually reported as the most dangerous.

Chapter Conclusion: It would appear that Shadow People are real, have some sort of capabilities and knowledge that probably exceeds our own. Furthermore, if they can move about invisibly, then nothing we have can be private to them. They can look through anything we have in our homes, at work, or anywhere, even in government secret facilities. They are here! Moreover, they do have an impact on us, one that raises our fears, and perhaps rightfully so. Imagine for example, being in the privacy of your own home, thinking your alone, but there is some stranger or "something" hidden from your sight, but perhaps hovering right near you, whether you are in bed, the bathroom, or wherever. Not a very settling thought.

Chapter 14—What Are the Origins of Shadow People

Another photo of a real Shadow Person?

Now that we've established that Shadow People could be a very real phenomenon and probably are, then the next step is to try to figure out why they are here, what is there origin? Well, let's consider some of the theories other researchers have promulgated.

1. Shadow People aren't real. They are just an optical illusion or perhaps just hallucinations. Well, we've pretty much discounted that idea. The available evidence strongly suggests they are real. Something seems to be "there."

2. They are from "elsewhere" as in being extraterrestrial. This one is not so easily ruled out, and in fact, might be the correct explanation. Researchers, myself included, have long stated that extraterrestrials are here and have been here for a long time, and also seem to have advanced technologies. Shadow People could just be well-cloaked aliens, if you will, so that they can move freely among us and in our homes, businesses, and supposedly secret government military bases. This is a very plausible explanation for them being here. It would also explain our history of interactions with Shadow People over the centuries, as being UFO folk, they may have discovered us humans long ago.

3. Shadow People are from another dimension. This explanation of their origin is also hard to dismiss. These could be interdimensional beings, and this would explain their Shadow-like appearances and ability to appear and seemingly disappear at will. So, the idea they may be beings from another dimension cannot be ruled out.

4. **Shadow People are time travelers.** This could possibly explain their origin. Moreover, the rules of time travel may forbid them actually being here as solid individuals, in which case, what we see are projections sent to us through time. The fact they can have some physical impact on us, if only indirectly, would also be explained in that such beings might be able to exercise some small amount of control on their surroundings, even as a time traveler. In other words, they may only partially exist in our time, but enough so to occasionally manifest an impact on objects around them to some degree.

Are Shadow People Time Travelers?

5. Shadow People are specters or ghosts of some sort. I discount this last idea almost completely. First, we don't even know if ghosts exist, and if they do, why would they appear as just shadows, and repeat certain types of shapes, like that of the "Hat Man?" Moreover, I doubt ghosts could use any sort of vehicles or such types of devices. Do I rule out the idea that Shadow People are just a type of ghost?" I do not, but I do discount it as the most likely of explanations for their existence.

6. Shadow People are part of a greater reality; one we cannot usually see or sense. In this respect, we think of Shadow People as being metaphysical in nature, but not being supernatural beings, such as ghosts. In other words, they are a natural part of our reality, one we just aren't privy to because of our own physical limitations, such as those of sight, hearing, smelling, and touch. Pets seem to be more aware of them in some instances. And if not ghosts, could they be demons or angels? Possibly. Perhaps people thinking they have a guardian angel is just because, on some level, they feel a presence often near them when there is nobody in sight. Are they demonic? Again, possibly, but I just don't feel that to be true. Why would demons appear the way they do? What could they hope to achieve? Not to mention this also means one has to believe in a god, heaven, and hell where such creatures might reside.

It is far mor likely they are just a being, different from us, but not necessarily either good or evil in the biblical sense, anymore than a bear, wolf, or snake is evil in such a way. However, the fact that they do seem to evoke a fear in us, even though they don't seem to attack us, is worrying. Again, perhaps this is just our response to the unknown, or maybe we sense them on a subliminal level and when actually confronted with them, our fear instincts just naturally kick in.

With regard to this last, there was the case where a woman was in her apartment in the living room. A short corridor connected the room to the front door and also the kitchen entrance, which was right off the foyer there. Her dog started barking loudly as it stared down the hallway toward the kitchen door and front door.

Then a shadow leaned out from the kitchen doorway. The person screamed, and the dog barked furiously. When the tenant ran to the kitchen, where the shadow had ducked back into that room, she saw nothing. With the lights on, the small galley kitchen was empty, not a sign of the shadow. There were no other exits from the kitchen. This is yet another example of their capabilities, and the video of this encounter is quite compelling. This video is available as part of a collection of such videos at the Slapped Ham YouTube channel.

Although, many legends through history of ghosts, ghouls, banshees, elves, goblins, trolls, fairies and such might just stem from past encounters with these shadow folk of a "greater reality," it doesn't tell us what they want.

7. Shadow People may be projections. There is a real possibility that Shadow People are actually sort of projections of people having out of body experiences, in that we could be seeing astral projections of people Is this likely? The truth is nobody knows, but some people do claim and rather convincingly so, that they have out of body experiences and can see, hear, and move about usually unnoticed in different places they travel to. Some even claim they can physically influence their surroundings.

Chapter Conclusion. What can we conclude about why they are here? Well, they could be a natural part of our existence, a "greater reality" that we simply aren't attuned to. In which case, they may have existed here on Earth as long as we have. Alternatively, they might be from elsewhere, either another star system, or another dimension, or even another time. The important thing is that wherever they come from, they are now here! So why are they here? Do they have an agenda? And if so, what does that bode for us? And again, think about this: if they can move in out among us and our homes and buildings, then we may never be completely alone, and we may have "something" watching us and our children, even when we think we are completely along and so safe.

Chapter 15—What Do They Want?

Shadow Person on a Bridge?

Why are Shadow People here? What do they Want? Well, the list could be endless, but here is what we know, or at least think we know, but remember, what the "DON'T do, is as important to us as what they "DO" choose to do:"

1. They seem to really be here.

2. They seem to have been here for a long time.

3. They either have cloaking devices or they can actually choose to appear and disappear, and even walk through walls seemingly at will. Either way, they seem to be able to control most instances of when we can actually observe them, unless of course, we only see them when they make a mistake and/or because of a malfunctioning cloaking device.

4. They don't seem to like light at all. They are truly creatures of the shadows in that regard. They appear mostly at night, or indoors, in darkened rooms, basements, tunnels, abandoned buildings (again, mostly appearing at night), etc.

5. They don't steal anything, at least, not in the vast majority of reports of them. Therefore, it seems they don't consider anything we own as valuable or important to them.

6. Generally speaking, with a few exceptions, they don't seem to physically harm people, unless people dying in their sleep is the work of such beings, as people in the past thought. However, they do seem to touch people at times, lift locks of their hair, as with women, take blankets off of sleeping people, and even reaching up into baby cribs. They also have been videoed sneaking up under the covers of people in bed, and in so doing, actually raise the covers of the bed while on camera. Why they do this is a mystery, but so far, people don't seem to be dying too often from these interactions, unless something like SIDS (Sudden Infant Death Syndrome) or again, deaths in our sleep are caused by them.

What we suspect. That's about the extent of what we know, and here is what we just think we know, or suspect about them:

1. They do seem to have an almost symbiotic or perhaps parasitic relationship with us humans. By appearing in our dwellings and other buildings and fleetingly touching our hair or bodies in some way. Perhaps they are just fond of us, or are curious as to who and what we are?

2. Nature of their business with us. Whatever the reason, they are here, whether because they are just another species on Earth (admittedly a weird-seeming on to us, if so), or they have some sort of business with us. The nature of just what that business could be, can run the spectrum from just being playful entities (like many poltergeists), to creatures that are gathering immense amounts of information on us, and/or keeping a close observation on every aspect of our lives for some reason. If other words, if it's not our things they are interested in, then it must be us!

This interest also could be for a number of reasons, and these range from just idle curiosity on their part, to planning something big (a coming invasion if they are extraterrestrials?) at some future point and so need all the data on us they can get. There are even reports of Shadow People invading people's dreams, as if to exercise control of us mentally, or manipulation of us through such dreams.

But whatever the reason, they are here, and they do interact with us, and it is imperative we find out why and what their motives might be. Are they stealing our energy or "baby's breath" as one person put it, or are they after something more than just grazing on us, as if we were so many cattle?

Chapter 16--Measures and Coping Strategies

To be honest, since we know so little about what makes them tick, so to speak, it becomes very hard to know how to deal with them. But there are some things that seem to work well, and others I am more skeptical:

1. Turning on lots of lights. This is probably the single best thing you can do. By having a lot of lighting on, especially at night, this seems to keep Shadow People mostly at bay. They seem to be able to withstand bright light but only for very brief periods, mere seconds. and then they flee to darkness. This seems to hold true even in abandoned buildings. When a flashlight is shone at them, they shrink back. So, your best defense seems to switch on a light and flood the area with it. Dispel the darkness and it seems you dispel the Shadow People, in only temporarily why the light is on.

2. Sage. Some people swear by that burning sage on the premises not only drives out evil spirits and demons, but Shadow People, as well. Maybe this works, and maybe it doesn't. It's possible that the smell is highly repugnant to Shadow People, and so it might drive them away. I am rather skeptical of this approach because it is totally unproven as being of any help so far, at least in an empirical data sense. If Shadow People are real living entities than foul-smelling burnt herbs may or may not banish them. In either case, I need some kind of real proof that this works to convince me of this method's effectiveness.

3. Performing rituals. White (Wiccan) style magic, or otherwise, would also seem to fall into the sage use category as far as I'm concerned. I simply question whether performing a so-called "magic" ritual will do anything about anything. It wouldn't drive me away, and I doubt it would drive away Shadow People. They might even get a good laugh out of such an attempt, if they can laugh…. This is not to denigrate people who believe in such things, but I would need far more evidence than I've seen so far (none, as it turns out) to place any real faith in the use of such rituals to drive out Shadow People.

4. Religious symbols and invoking Jesus. For the same reasons as Number 3, above, I don't really place much faith in this approach (no pun intended) to do much either. The evidence suggests strongly Shadow People are real. Therefore, they probably are not demons, evil spirits, or spiteful ghosts or poltergeists. They would not be of such a supernatural nature if they naturally belong to our reality.

Only religious believers, believers in Jesus and such, seem to attempt this approach and it seems to be more because of their beliefs and faith than of any real practical help. However, when all else fails, one can always try praying about such matters, but as to whether this will do any real good, other than making the person who is doing the praying feel better, is doubtful. Praying for a bear to go away when it's attacking you, probably wouldn't help much, and I don't think it helps much here, either. However, I leave it to the individuals involved to try such an approach, and I truly hope (although I seriously doubt it) that it will help ward off such beings. Although I don't personally place much faith in faith, perhaps it works for those who do.

5. **Seeking professional help**. Looking for some sort of professional support would, again, probably be of limited use, as well, under these circumstances. Sightings of Shadow People can be few and far between for any given individual, and getting such entities to repeat such events so as to provide proof they exist to professionals, probably won't work most of the time. So, frankly, the "professional," whoever that is, is more likely to focus on the viewers mental state and emotional health to find some reason for such

sightings. ("You are overworked, stressed, and you imagination is seeking a way for you to mentally escape," or "you are drinking or using some type of psychedelic drug," and so on). Shadow People don't seem to be a repeatable experience in the sense that the viewer has any control over when and how such things appear. They simply cannot be conjured up on demand.

6. Mindfulness sessions. Meditation and self-care can help a person to get through the repercussions of a Shadow Person sighting, but I doubt if they would affect how often or not such an individual might see such apparitions. Furthermore, such steps might make the victim feel better, but I haven't heard of any evidence at all that they can banish Shadow People.

Chapter Conclusion. Although, one can always try to approach such sightings in a certain prepared mental state to change the outcomes, I haven't heard that this makes any difference as to how often, or to what extent, people have these Shadow People sightings. Although one can use meditations, prayer, mindfulness and self-care to deal with the unsettled consequences such sightings can cause, they seem to have little or no impact on whether a person witnesses such things in the first place. The same seems to hold true for using any of the other methods listed above, with the single exception being the use of light to fend off Shadow People.

Conclusion

So, what have we come up with from all of this discussion about Shadow People? What can we derive from all this, if anything?

Well, the evidence strongly suggests they actually exist. The evidence would also seem to indicate they are among us, and maybe even far more so than with realize. They are in our houses, at least at times, apartments, or whatever. They are seen in factories, warehouses, hospitals, and even police stations. In short, they seem to be everywhere. Furthermore, are we ever really alone, or is one of them (or more) always lurking much closer to us than we think?

The also seem capable of using cloaking devices in all sorts of ways. Either this, or whatever these beings are, they can transcend the laws of physics as we know them. No matter with explanation is true, if so, this is a frightening thought; that they can do all sorts of things we cannot, whether through the use of technology or just naturally. This puts us at a strong disadvantage in any attempt to combat them, if true.

Moreover, we aren't at all sure where they came from, but it seems, judging by historical information, they may have been here a very long time, or perhaps have always been here without us realizing it. We also aren't at all sure why they are here, but the constant feelings of fear and dread they invoke in us does seem to indicate their presence may not be a good thing.

Why do they invoke such fear in us? Is it just a defense mechanism on their part, that they can project such an emotion into us, or is there a real, or at least, more tangible reason why we should fear them? They might well prove to be a danger to us. With them, being an unknown quantity, that idea has to be considered and seriously. Moreover, if they aren't a symbiotic species with us, then they might well be a parasitical one. They might need us for some reason. And that reason may be a dark one.... Reports throughout history of babies dying in cradles and cribs, people dying in their sleep, and then seeing videos of these things creeping into people's bedrooms, and beds, is spooky stuff!

Of course, there is no smoking gun with regard to whether Shadow People steal our energy, breath, or lives from us, but it is a very odd coincidence, even so that this happens in our bedrooms and those "things" find bedrooms to be one of their favorite places to be. This "coincidence" should be investigated, at the very least.

Whatever the visitations from Shadow People portend, this is one mystery that will probably be an ongoing one for some time to come. And we, the apparently helpless victims of such encounters, must simply put up with them and somehow endure, as we have endured for centuries so far. But the ongoing mystery of Shadow People is going to be with us for some time to come, I fear.

One thing we can do is open more dialogues with others of us, find people willing to research this phenomenon, and attempt to find real answers that just might help us survive as a species. After all, we Homo Sapiens have managed to somehow be the only human species to survive, and it seems we may have been in competition with many other such species over the years, only to ultimately beat them at the evolutionary game of survival.

If Shadow People are just a different sort of intelligent species, then we may have unknowingly been in competition with them, as well, throughout our history. If so, this begs the question as to which species will ultimately survive in the long term. Granted we have survived so far, but are we being manipulated and controlled by these entities? Is that why they pry into our lives and homes so thoroughly, to help maintain that control? Are we, in some hidden way, subservient to these beings? Are they "feeding" off of us?

It is a curious thing that we've developed the myth of vampires over the last centuries. What's even more curious is that the description of vampires is a close match for that of Shadow People.

1. Both are creatures of the dark.

2. Both shun the light of day.

3. Both seem to have supernatural or extradentary powers to vanish quickly.

4. Both could well be "feeding" off of people, and if too much of this goes on, the person can die.

Again, interesting how alike the descriptions of the two things are, vampires and Shadow People. Perhaps, a little too coincidental, maybe? Food for thought. In any case, we have to know the answers to these questions if we hope to survive as a species. Ignorance, to paraphrase, in this case would not be "bliss." We need to know!

Will that help us survive? Well, there are no guarantees in this universe, but knowledge helps. It is a tool and a weapon, if necessary, and when battling the unknown, it's important to have such, if needed. Remember; this very well may be about our long-term survival as a species.

Final Thoughts. So, what do I personally believe about Shadow People? Well, I think they probably exist. I'm not one hundred percent positive, but let's say I am very close to that. Whatever Shadow People are, they are not like us, apparently, and they do make us fearful. They make me fearful. Yes, upon researching the facts, I do think they definitely could be a real threat, and we should keep that well in mind until we know for sure, otherwise.

In any case, when it comes to not wanting to see Shadow People, light is your best weapon as a defense. They do not seem to like brilliantly lit places, and prefer to stick to the shadows, or darkened areas, or simply preferring the night. Avoid these places and conditions if you are alone. There also seems to be some safety in numbers in that regard. As for all the other possible "remedies," you can try them, but proof of their effectiveness is severely lacking so far.

Most of all, we have to find out just what Shadow People are and what their motivations might be. Until we know that, we are just as much in the dark as they seem to be. And since they seem to know every little detail about us, totally invade our privacy to do that, I think it only fair that we get to learn a lot more about them, too!

Resources and References, Live Links
(At Time of Publishing)

https://bloody-disgusting.com/editorials/3788890/shadows-of-ourselves-the-unexplained-phenomena-of-shadow-people-dead-time/

https://villains.fandom.com/wiki/Shadow_People

https://www.liveabout.com/shadow-people-2596772

https://discover.hubpages.com/religion-philosophy/Shadow-People-Who-Are-They

https://medium.com/@emmalovesharry10/top-ten-shadow-people-encounters-from-around-the-world-mysterious-and-chilling-stories-b6858dd66471

https://www.psychologytoday.com/au/blog/shadow-boxing/201307/shadow-people

https://homespunhaints.com/shadow-people-in-peripheral-vision

https://www.shadowpeople.org

https://freaked.com/photos-of-shadow-people/

https://connectparanormal.net/2024/05/25/shadow-people-uncovering-the-elusive-entities/

https://www.monstropedia.org/index.php?title=Shadow_people

https://brickthology.com/2020/10/25/shadow-people/

https://darksiremag.wordpress.com/2021/04/23/reality-meets-fiction-shadow-people/

https://www.youtube.com/watch?v=dOTjSrvnEdA

https://www.science.org/content/article/illuminating-shadow-people

https://discover.hubpages.com/religion-philosophy/Shadow-People-What-Are-They-Where-Do-They-Come-From-And-Shadow-People-Stories

https://cphswolfpack.com/opinion/the-shadows-have-eyes/

https://toko-pa.com/2019/08/20/shadow-dreams/

https://www.nataliakuna.com/shadow-people--dark-beings.html

https://www.youtube.com/watch?v=4Xy64asE5RY

https://www.amazon.com.au/Walk-Shadows-Complete-Shadow-People/dp/173391935X

https://www.nature.com/articles/443287a

https://skeptoid.com/episodes/4175

https://www.youtube.com/watch?v=1onU1JSpVII

https://www.youtube.com/watch?v=1onU1JSpVII

https://occult-world.com/shadow-people/

https://en.wikipedia.org/wiki/Shadow_person

Slapped Ham Videos Channel:

https://www.youtube.com/channel/UCw4ccFtBN7dhQBcHmE0qylg

Shadow People, as imagined by an A.I.

A.I. Rendition of Djinn Shadow Person

Djinn or "genies" as imagined by an A.I. and based on descriptions.

Author's Biography

Rob Shelsky is an avid and eclectic writer and averages about 4,000 words a day. Rob, with a degree in science, has written a large number of factual articles for the former AlienSkin Magazine, as well as for other magazines, such as Doorways, Midnight Street (U.K.), Internet Review of Science Fiction (IROSF), and many others. While at AlienSkin Magazine, a resident columnist there for seven years, Rob did a number of investigative articles, including some concerning the paranormal, as well as columns about UFOs, including interviews of those who have had encounters with them. Rob also writes fiction, including science fiction, horror, fantasy, and paranormal. His trilogy, the Apocrypha is currently with Permuted Press.

Rob has been interviewed on a large number of shows, including George Noory's Coast-To-Coast AM Radio show, House of Mystery, The Kevin Cook Show, Art Bell's Midnight in The Desert, The Warren XChange, Mysterious Radio, and many others. He has often and over a long period, explored the alien and UFO question and has made investigative trips to research such UFO hotspot areas as Pine Bush, New York, Gulf Breeze, Florida, and other such regions, including Brown Mountain, North Carolina, known, for the infamous "Brown Mountain Lights," as well as having investigated numerous places known for paranormal activity.

He has traveled abroad to do this, as well, as with traveling to sites in the United Kingdom, Canada, and other countries where UFOs have been reported. Rob was a member of MUFON, and a Field Investigator for this group. The author was even invited to speak at the Library of Congress, Washington, D.C. The author has many books on the various topics of UFOs, paranormal phenomena, and extraterrestrials. Rob was also the on-camera expert for a new UFO show in development for the Travel Chanel through Karga 7 Studios in Los Angeles

Rob's fiction credentials are extensive. Combined with his nonfiction books, he has a total of some 78 books published. These include, science fiction (thrillers), paranormal novels, including fantasy, horror, and more. For links to Rob's other books on the subject of UFOs, Time Travel, the Mandela Effect and other topics, please go to Amazon Kindle, Smashwords, B&N, etc.

www.ingramcontent.com/pod-product-compliance
Lightning Source LLC
Chambersburg PA
CBHW052325220526
45472CB00001B/281